I0509982

Plumbing
'Estimator'

PLUMBING
'ESTIMATOR'

Purchase THIS Book to save $'s
contact via email for free estimates - details inside

Author Sherman Turner

Plumbing
'Estimator'

Author Sherman Turner

'Plumber'
Apprenticeship

Copyright © 2020 Sherman Turner
All rights reserved. No part of this document may be
reproduced or transmitted in any form or by any means,
electronic, mechanical, photocopying, recording, or
otherwise, without prior written permission of and by
Sherman Turner.

The Table of Contents

Dedication.. vii

About The Book ix

Plumbing Answers Part One............................ xv

    Chapter #1 Lavs & Vanity Sinks.....................1

    Chapter #2 Kitchen Sinks..............................4

    Chapter #3 Toilets..6

    Chapter #4 Water..11

    Chapter #5 Hot Water Tanks 15

    Chapter #6 Sewers and Drains 21

    Chapter #7 Sump Pumps..............................27

    Chapter #8 Plumbing Answers.....................30

Plumber 'Speaks' Q and A Part Two 49

    Plumber 'Speaks' Q and A...........................51

Plumber 'Speaks' Plumbing Tips Part Three ...67

    Plumbing Tips 69

Budget Cost Estimating 83

    Chapter #9 Budget Cost Estimating.............85

Free Estimate and Evaluations......................113

vi

Dedication

In Memory of *Malcolm X*

This book is **dedicated** to my loving daughter *Sabrinna L. Turner* my one and only loving daughter.

viii

About The Book

Author and *Master / Plumber* Sherman Turner wrote honoring an honorable man who dedicated his life helping mankind and his brothers. This book *(Plumbing* *'Estimator')* depicts 1860s and 1960s how unjust **"Jim Crow" laws** still existed in United States of America to hold minorities bondage until Malcolm X, spoke-*(Plumbing* *'Estimator')* saves $10,000 in savings while providing *"Free Plumbing Answers"* and *"Free Estimating Services"* and charts. Learn do's and don'ts in Plumbing and start saving. Don't hesitate, don't wait, and don't be late! This is one-of-a-kind book and your best-buy book!

Why did I write this book? As a minority person and poor person, I know the struggles of getting correct information and not having the knowledge to know the **do's** and **don'ts** in Plumbing?

ix

Master-Plumber and Author has more than 40 years' experience and more than 30 years' experience as a Senior Plumbing Estimator.

Thank you, for allowing me to be of any type plumbing service and help for your *"Free Plumbing Estimate Evaluations."*

This *"Free Estimate Evaluation Service"* has never been offered before, not in books, not on TV, not on radio and not on internet.

A *"Free Estimate Evaluation"* is an appraisal of something to determine its worth or fitness. In most cases this eliminates Plumbing Contractor rip-offs!

Plus, *"Free Estimate Evaluation"* is an appraisal which recommends you make and stay within your budget or goals. This is very helpful for many Job Projects and Jobs!

This *"Free Estimate Evaluation Service"* recommended for disabled, veterans, sick and the elderly. To avoid getting ripped-off again!

(Special) For same day 24-hour bids!

For **(SB)** Small Businesses and **(PC)** Plumbing Contractors: seeking win more competitive bids it is recommended to have an experienced estimate reviewer, review your ***"Bid Summary!"*** (1 - 4 pages) To make your bid a competitive winner!

(Special) For same day 24-hour bids!

For **homeowners** and all **others,** I recommend an estimate budget evaluation cost! Because if you feel like Plumber price is too high, nine times out of ten, the Plumber's price is too high!

Most **Plumbers** and **Contractors** rip you off for ten times as much as the original cost. Fight back get an estimate budget evaluation cost!

(If you don't know what to do or are you seeking Plumbing advice?)

Go to: ***"Free Estimate and Evaluations."*** Our answers are guaranteed to start saving your money funds.

(Special) For same day 24-hour bids!

Go to: *"Free Estimate and Evaluations."* Our answers are guaranteed to start saving your money funds.

Additionally, use free *"Estimating"* system with computer charts for checking labor costs and material costs too! This book explains the do's
and don'ts. Most books say, "do this" but never say *"don't do that."*

xiv

Plumbing Answers Part One

xvi

Chapter #1

Lavs & Vanity Sinks

What does *"Lav"* mean in plumbing?

Lavatory ("Lav") fixed bowl or basin with running water and drainage for washing. Many use the word "lavatory" meaning a bathroom.

What is a standard size bathroom sink?

There's no standard size for a bath sink. Most round sinks are 16 to 20 inches in diameter, and most rectangular sinks are 19 to 24 inches wide and 16 to 23 inches front to back.

Bathroom Sink Styles:

Pedestal Sink, Wall-Mounted Sink, Vanity Sink.

PLUMBING 'ESTIMATOR'

Pedestal Sink

Single Bowl Sink
Stainless Steel

Is vitreous china better than ceramic?
Vitreous china is better at resisting spills, scrapes, or other bathroom mishaps. The high

SHERMAN TURNER

gloss enamel is very durable and creates a stain-resistant surface.

Bathroom Sink
Wall Hung

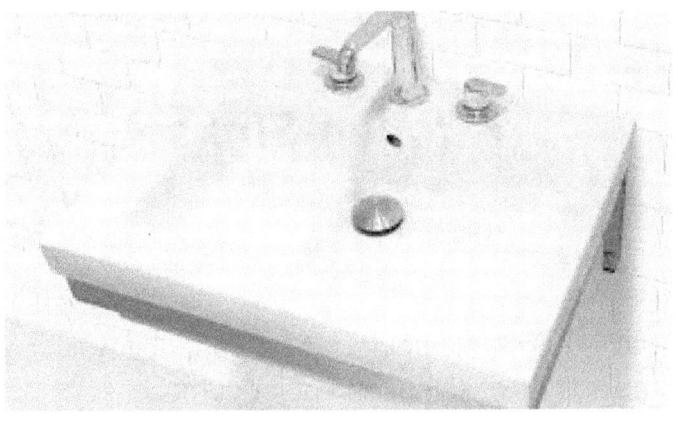

Vanity Sinks

PLUMBING 'ESTIMATOR'

Chapter #2

Kitchen Sinks

What is a sink in kitchen?

Kitchen Sink plumbing fixture used for washing hands, dishwashing, and other purposes.

Kitchen Sink "P" Trap
The curved pieces of drainpipe underneath your sink, commonly referred to as P-traps. Traps come in 1 1/4 inch (standard bathroom sink) or 1 1/2 inch (standard kitchen sink)

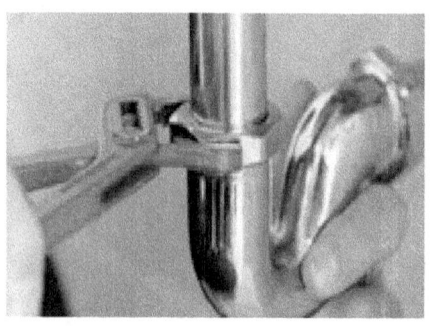

SHERMAN TURNER

Double Bowl Kitchen Sink

What kind of kitchen sinks are in style?

Self-Rimming Sinks,
Prep Sinks, Iron Island Sink:

PLUMBING 'ESTIMATOR'

Chapter #3

Toilets

Which type of toilet is the most common and affordable type?

The most common **type of toilet** in the U.S. is the two-piece toilet.

Most toilets are available in a 12-inch rough-in, which is the standard distance, but a 10- or 14-inch rough-in may be needed in some homes.

What are the types of toilet?
According to the bowl design toilets can be classified around four types:

- Round bowl toilet.
- Square bowl toilet.
- Elongated Bowl toilet.
- Rectangular bowl toilet.

Most toilets are available in a 12-inch rough-in, which is the standard distance, but a 10- or 14-inch rough-in may be needed in some homes.

Question: Why would a toilet bowl suddenly go dry?

A toilet like any fixture can lose its seal (water). By negative pressure or positive pressure or back siphonage.

Question: What is the best way to move the toilet flange a couple of feet and is there an adapter to go from cast iron to PVC?

Yes, No-Hub Band or Fernco Coupling is the preferred transition from cast iron to PVC plastic.

Question: What is the correct type seal for a toilet flange? Why should the toilet have this type seal?

Always use deep seal wax gasket. That type seals right inside and on top of the toilet flange, preventing sewer gases from entering the house.

Picture shows the correct placement of the plumbing deep seal wax gasket.

PLUMBING 'ESTIMATOR'

Note: Plumbers or handymen who don't seal toilet flange properly causes water leaks resulting in water damages, in thousands of dollars.

Question: What is the correct wax seal for a toilet flange that is flush with the floor?

Try Ultra Seal type wax gasket, I usually use the thickest wax ring I can find with the plastic funnel thing embedded in it.

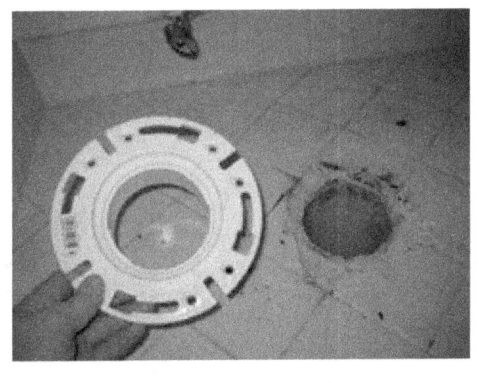

Note: The below picture illustrates different parts of the toilet that the Plumber must check is working properly.

PLUMBING 'ESTIMATOR'

Please see the most important floor seal at the closet flange. Make sure to check and test that your toilet is fasten securely and does not rock. Did the Plumber seal your toilet fixture space and the floor?

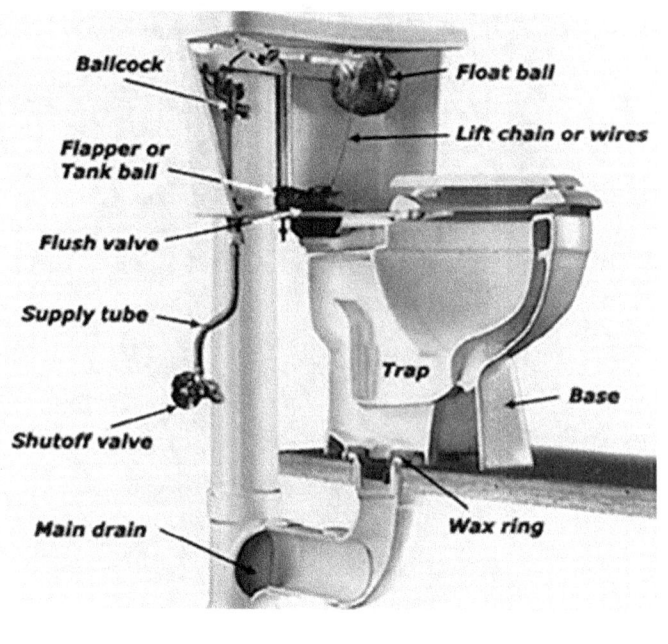

Chapter #4

Water

Whenever you have a water emergency in the house such as a leak or water causing water damages to property. TURN-OF the water!

Question: How can you increase the water pressure of a faucet?

Usually the aerator is plugged; but, sometimes the rubber inside the stops under the sink wear out and tear away.

Sometimes the supply lines get kinked and need to be replaced. If you get a flex tube to connect to your stop under the sink, you could attach it and blow the water into a bucket to see if you have pressure there. If there is pressure, then you know the problem is above the stops.

Question: What causes a loud knocking sound

PLUMBING 'ESTIMATOR'

like a jackhammer in the home plumbing when water is being shut off?

This is called "water hammer" and you need to install "water hammer arrestors" in the piping system where required.

Question: What is wrong when pipes make sound as you turn on the water?

It could be you have a loose washer on a faucet. If that is what you mean when turning on the water. Please install new washers and if the faucet become defective, then please buy and install new faucet.

Unscrupulous Contractors will say you need total new piping water replacements. Thanks for asking question, because just saved yourself plumbing bill worth many thousands of dollars.

Question: What causes or how to prevent water piping to rattle, shake and make loud noisy sounds? Does this mean major new water lines and new piping should be installed?

You should have Water Hammer Arrestors installed, same as the types and examples shown below. There are many Companies and

12

SHERMAN TURNER

Manufactures of Water Hammer Arrestors.

The water piping and water lines should be properly supported with the proper type of insulation as required.

PLUMBING 'ESTIMATOR'

Question: How can I increase water pressure at my kitchen sink Faucet?

When you start having reduced water pressure at your kitchen sink or any sink. This means you may have some blockage in the aerator.

Clean the aerator out or replace with a new one. Then turn your water back on and your water pressure should have increased adequately.

SHERMAN TURNER

Chapter #5

Hot Water Tanks

Hot Water Tanks:

Hot Water Tank is a convenient **heat storage** medium because it has a high specific **heat** capacity. Water is non-toxic and low cost. An efficiently insulated tank can retain stored heat for days, reducing fuel costs.

How do **Hot Water Tanks** work?

An **electric water heater** works essentially the same way as a **gas water heater**. It brings cold water in through the dip tube and heats it using the electric heating elements inside of the tank. The hot water rises in the tank and is moved throughout the home through the heat-out pipe.

What size **Hot Water Tank** do I need?

PLUMBING 'ESTIMATOR'

For 1 to 2 people: 30-40 gallons. For 2 to 3 people: 40-50 gallons. For 3 to 4 people: 50-60 gallons. For 5+ people: 60-80 gallons.

Hot Water Heaters:

Most demand **Hot Water Heaters** are rated for a variety of inlet temperatures. Typically, a 70ºF (39ºC) water temperature.

What brand of Hot Water Heater is the most reliable?

- Rinnai Water Heater. ...
- Ecosmart Tankless Water Heater.
- GE GeoSpring Water Heater. ...
- Stiebel Eltron Water Heater. ...
- Bosch Water Heater. ...
- Takagi. ...
- Kenmore Water Heater. ...
- American Standard Water Heater.
- Rheem

How much does a Plumber charge to install a **Hot Water Heater**?

Plumbers typically charge $75-$150 per hour and can typically install a **Hot Water Heater** in a day (6-8 labor hours), for a total labor **cost** of $600-$840.

SHERMAN TURNER

Why does the **hot water** run out?

The **sediment** buildup in your **Water Heater**.

This is the most likely reason your hot water is running out too quickly. It happens because water picks up natural minerals and sediment on its way to your Water Heater. Then, over time, the sediment sinks to the bottom of the Water Heater Tank because it's heavier than the water.

Can a **Water Heater** explode?

Yes, if the temperature is set too high or the pressure relief valve of a Water Heater malfunctions, a Water Heater can explode. Although it is unlikely for Water Heaters to explode, but when they do, they operate much the same as a rocket.

How do you know a **Hot Water Heater** is going bad?

Lack of Hot Water. Popping or Rumbling Noises. Water heater noises. Cloudy Water. .

Leaking or Faulty Pressure Relief Valve.

Leaking Hot Water Tank.

PLUMBING 'ESTIMATOR'

Electric Hot Water Tank

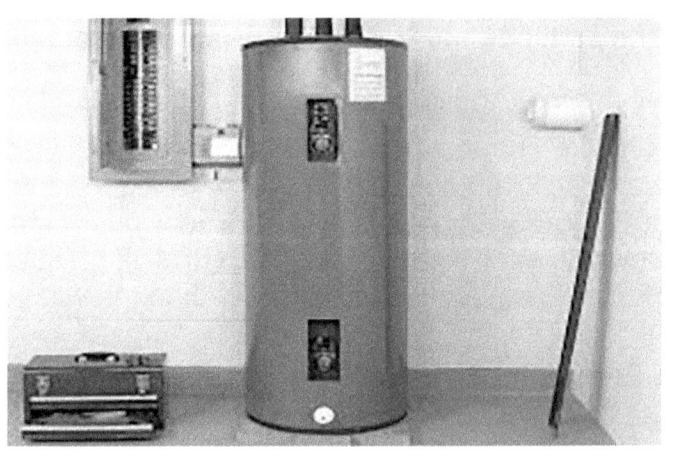

Gas Hot Water Heater

Question: How do you stop a noisy hot water heater from making noise?

You can reduce the noise by draining the Water Heater and removing the lime deposits the best you can.

Most likely you can't reduce the noise if the gas Water Heater and you have hard water. The only way to fix this is to buy a new Water Heater and install a Water Softener.

SHERMAN TURNER

Question: How do you stop pipes and water piping from sweating?

You need to wrap the pipes with insulation. Get insulation that the inside diameter of the insulation is larger than the outside diameter of the piping system.

Note: The wall thickness of the insulation should be at least **"one inch thick."** Make sure to use "duct tape" or "electrical tape" when covering turns in the piping system.

Question: Why would kitchen sink or washroom sinks have a stink smell?

The only reason I know that the stink smell would come about is that the P-Trap or U-Bend underneath a fixture type sink has lost its trap seal due to evaporation.
Refill the trap seal and pour water into the sinks to have a trap seal.

Question: What causes sink drains to make sucking sounds?

The sucking noise is a good thing, it means that the drain is working properly, and the noise comes from a swirl in the water going

PLUMBING 'ESTIMATOR'

counterclockwise and pulls the water. The noise is only air.

Chapter #6

Sewers and Drains

Question: When roughing in a new toilet what is the correct distance from the center of the drain to the wall?

A minimum of 18 inches off the finished side wall and a minimum of 12 inches of the finished back wall.

However, 18" side clearance is the international standard for wheelchair and assisted toilet maneuvering room. Maybe make more rooms comfortable for larger people.

Question: How do you make repairs when the Lavatory sink is plugged and not working?

Plumber shown below is using the correct tool for the fixture trap removal. Once the fixture trap is removed and cleaned the Lav sink will drain properly.

PLUMBING 'ESTIMATOR'

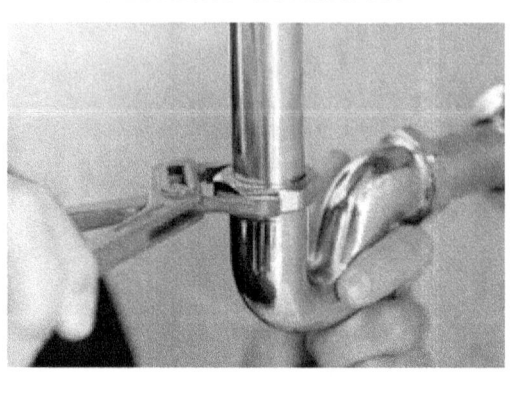

Question: What is waterproofing? Where should waterproofing be used?

Use waterproofing sealers and paints which are recommended materials for their usage only. However interior waterproofing sealers and paints will not alleviate all potential sources of basement leaking problem areas.

Question: How often should downspouts and gutters be cleaned?

Rain gutters and downspouts should be cleaned regularly. Remove all leaves, dirt and debris at least once or twice a year.

Question: What are "French Drains?" And where should they be used?

SHERMAN TURNER

"French Drains" are used to channel water out of the building which then would be pumped out of the building through means of a sump pump and sump pump pit.

Question: How often should roofs be inspected?

Roofing systems should be inspected at least once a year.

Question: Who best installs new roofs? A General Contractor or a Roofing Contractor?

A licensed Roofing Contractor who offers a guarantee for their work and materials.

Question: How long does a roto-rooter job last?

It depends on what's wrong with your sewer. It could take anywhere from 10 minutes to a whole day. Usually blockages in the main occur either from roots or breakages in the pipe.

If the line has been snaked and you still have a problem, I would recommend having the sewer, camera for pictures of the interior sewer system piping.

PLUMBING 'ESTIMATOR'

Question: How do you keep roots out of the main sewer line?

The only thing you can really do is snake out the sewer drainage system and get it as clear as possible. Then you can dump copper sulfate down the system.

It probably won't kill all the roots, but it stops them from growing. This works best if you do it in the spring during the growing season. And you should probably do it about twice a year.

Another chemical to use is "Root X" because "Cooper Sulfate" is illegal in some states.

Putting any kind of chemicals down your sewer drain line is illegal in most countries around the world; it is considered an environmental hazard. If you are caught by authorities, you may be in serious trouble and the implications will be severe.

Therefore, it may be best to avoid using chemicals and rather rod or snake your sewer line, then cut down brushes or trees suspected to be growing over your sewer. Or replace your sewer line and maintain your sewer line by rodding it every six months.

SHERMAN TURNER

Question: What is a plumber's snake used for?

It is a tool they use to shove down your pipes to clean them out if they are clogged and the snake can be either electric powered or by hand. See the example sewer machines below:

Electric Sewer Machines:

PLUMBING 'ESTIMATOR'

Electric Sewer Machines:

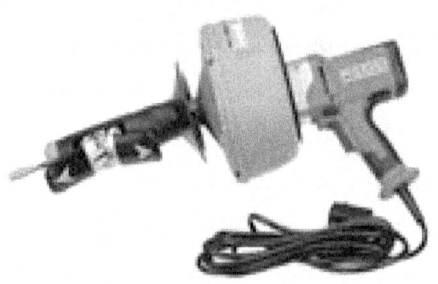

SHERMAN TURNER

Chapter #7

Sump Pumps

Question: Every time it rains my basement gets flooded from water under basement floors and walls, what can I do about this situation and problem?

Seems like you may need a sump pump and basin used to manage surface runoff water and from underground water aquifers.

PLUMBING 'ESTIMATOR'

Question: How do you measure the water evaporation from your swimming pool?

Measuring Water Evaporation First; mark the water level on the wall of the pool. After some time, mark the water level again. It should be lower than the original one.

Question: What would cause water to seep out from under the toilet bowl onto the floor when the toilet is flushed?

You may have a partial clog somewhere in your pipes; therefore, the water is backing up in the first place.

Or the wax seal under the bottom of your toilet (you can't see it because it's UNDER the toilet) is broken.

Question: What causes the popping or tinkling sounds as hot metal pipes cool off?

Contraction, though the sounds are usually noisier when pipes are expanding as hot water runs through them. The sound you hear is caused by the expansion and contraction of the metal cause by the heating and cooling.

SHERMAN TURNER

Question: What would cause the hot water to be rusty and brown?

Sometimes chemicals cause rust to get into the water lines. Or change in water pressure will cause this rust to turn loose and come through into the bathtub or other appliances.

If only the hot water is rusty! It's 100% your hot water tank!

The real reason why you're not seeing it in your cold water is because your cold-water lines run directly to your water fixtures.

Question: Who uses concrete roofs?

It is very common to find commercial and residential buildings having concrete roofing.

PLUMBING 'ESTIMATOR'

Chapter #8

Plumbing Answers

Faucets and drains are the parts of your plumbing system that are most likely to break down, so therefore you will need repairs more often. Faucets leak or drip and drains get clogged up.

Toilets: Clogged toilets are the most common plumbing problem can have. If a toilet overflows or flushes sluggishly clear the clog or backup with a plunger or closet auger.

A recurring puddle of water or water leak recurring on the floor or around the toilet it may be caused by a crack in the base of the toilet. This persistent puddle problem could possibly be from the toilet tank.

SHERMAN TURNER

Toilet stability the toilet fixture shakes or rocks when it is not fastened securely. Check the closet flange and secure all loose connections including replacement of the closet wax ring gasket with nuts and bolts, too.

Water on the floor around toilet - check if the toilet base in the toilet tank is cracked and leaking. Sometimes you may have to insulate the toilet tank to prevent condensation.

Sink drains: every sink has a drain trap and a fixture drain. A sink gets clogs or plug ups by a buildup of soap and hair in the trap or fixture drain line. Remove clogs and plug ups by using a plunger, disconnecting and cleaning the trap.

PLUMBING 'ESTIMATOR'

Clogged and plugged lavatory sinks can be cleared with a plunger.

Dump out the debris after removing the trap. You may need a small wire brush when cleaning the trap bend. Reinstall the trap bend.

Faucets eventually just about all faucets develop leaks in drips.

If your old faucet continues to leak after repairs are made, then you know that you will have to replace the old faucet with a new faucet.

SHERMAN TURNER

Water pressure at the spout of the faucet is problematic water pressure seems low water flow is partially blocked. Clean the faucet. This is this is not correct the situation, take a closer look under the sink if the pipe in his old galvanized piping, replace the corroded galvanized pipes with copper or plastic new water piping.

A dishwasher that is past it's may be inefficient in many ways. It probably was not designed to be very efficient to begin with. However, most significantly, if it no longer cleans effectively, you are probably spending a lot of time and how water pre-rinsing the dishes.

PLUMBING 'ESTIMATOR'

This alone can consume more energy and water in a complete war cycle on your machine. Therefore, even if the old dishwasher still runs, replacing it with an efficient model can be a good investment upgrade.

In terms of size and utility hookups, dishwashers are generally quite standard if your old machine is a built-in and your countertops and cabinets are standard sizes.

Most full-size dishwashers will fit right in. Of course, you should always measure the dimensions of the old unit before shopping for the new one to avoid an unpleasant surprise it installation time. Also, be sure to **review the manufacturer's instructions** before starting any work and installations.

SHERMAN TURNER

Replacing an old, inefficient dishwasher is a straightforward project usually takes just a few hours. The energy savings begins with the first bowl of dishes and continues with every load thereafter. Therefore, we always recommend using a newer model of the appliance.

Food disposers are standard equipment in the modern home and most of us have come to depend on them to macerate our plate leavings, so our crumbs can exit the house along with wastewater from the sink drain.

If your existing disposer needs replacing, you will find that the job is relatively simple, especially if you select a replacement appliance that is the same model as your old one. In that

PLUMBING 'ESTIMATOR'

case, you can probably reuse the existing mountain assembly, drain Steve, and drain plumbing.

Most food disposers are classified as are classified as "continuous feed" because they can only operate with an on/off switch on the wall is being actively held down. Let go of the switch and this disposer stops.

Icemakers, the most expensive refrigerators come with icemakers as standard equipment, and practically every model features them as an option (a refrigerator with an icemaker usually calls about $100 more).

Most icemakers come either preinstalled or are purchased as an accessory when you buy your **new refrigerator**. An icemaker receives its supply of water for making cubes to through a ¼ inch copper supply line that runs from the icemaker to a water pipe. The supply line runs through a valve in the **refrigerator**, and it is controlled by the solenoid valve that monitors the water supply and sends the water into the icemaker itself, where it is turned into ice cubes.

36

SHERMAN TURNER

Refrigerator with icemaker

Automatic icemakers are simple to install is long as your refrigerator is icemaker ready. Make sure to buy the correct model for your appliance and try doing careful installation work. Icemakers water supply lines are very common sources we water leaks occur quite often.

Standpipe drains in many houses the washing machine drain holes is followed loosely over the side of the **utility sink** this arrangement is now frowned upon by the building codes.

PLUMBING 'ESTIMATOR'

Instead of hanging it over the **utility sink**, you should install standpipe that allows the **washing machine** to drain directly into the **utility sinks** drain line piping system.

A 2-inch standpipe is required by most codes, with a 2-inch P trap. The top of the standpipe should be higher than the highest water level in the **washing machine**.

Water softeners if your house has hot water running through his pipes, you have got a couple problems. Not only does your water do a poor job of dissolving soap, you also have plenty of scale deposits on your dishes, plumbing fixtures, and the inside of your Hot Water Tank.

SHERMAN TURNER

Water softeners fix these problems by chemically removing the calcium and magnesium that are responsible for the hard water.

Water softeners are installed after the water meter but before the water line branches off to the appliances or fixtures, with one exception. Piping to outside faucets should branch off the mainline before the softener because treating outside water is a waste of money.

Sump pumps, price ranges from about $100-$500 or more, depending on the quality and the features of the sump pump. Below is a picture of a standard sump pump.

PLUMBING 'ESTIMATOR'

First decide between a pedestal and a submersible pump. The standard pump, non-submersible, as shown below:

Submersible pumps sit in the water a good deal of the time, they have a lifespan from 5 to 10 years. However, most manufacturers offer limited 1 to 5-year warranties. Sump pump is measured by horsepower.

It is better to buy a cast-iron sump pumps, which last longer than plastic or iron types because of corrosion. Make sure the power cord is long enough because electrical extension cords are not to be used on sump pumps.

SHERMAN TURNER

Always place pavers or bricks underneath the sump pump so that mud, dirt, and grit not to plug up the operation of the pump.

Using a plunger or a "plumber's plunger" which is about 10 to 12 feet long, to push the obstruction down thru F.A.I. and hose trap, to clear the clogged or plugged up sewer line.

For some hot water down the line and then apply the pressure forced by plunging several times. The pressure from the simple tool can generate pressure to blow out obstructions quite quickly.

Using an **electric snake** if the obstruction is in the form of tree roots, instead of common obstructions like toilet paper rags leavings, and

PLUMBING 'ESTIMATOR'

garbage. You need to use a snake tool, which is normally used by plumbers and available in the hardware stores, for a rental fee.

In cases of extreme clogging and plug ups, call a professional plumbing company may use a jet hydro-vac machine.

A clogged sewer is a problem, which needs immediate attention as it can become a nasty nightmare if not taking care of in time.

The **below** is the type electric sewer machine or **electric snake machine** that you will need:

SHERMAN TURNER

Hot Water Heater problems normally become self-evident. The hot water faucet fails to summon or get hot water from the spout, you see scribbling or puddles near the Hot Water Heater, or the tank emits strange gurgling or pop and sounds coming from the Hot Water Heater.

Hot Water Heaters makes strange noises expanding and contracting metal parts, or more likely, minerals and higher water scale accumulations inside the tank can cause the noises coming from the Hot Water Heater.

PLUMBING 'ESTIMATOR'

To avoid scale buildups, every few months, open the **drain valve** at the base of the tank, and flush the tank until the rest runs out and you see clearwater.

Most Hot Water Heaters manufacturers recommend draining and flushing your hot water tank once per year or every six months and heart water areas.

This helps remove sediment and minerals that collect at the bottom of the tank. Sometimes the sediments chunks may be too large to pass through the drain valve on the tank.

If you hear a boiling sound of water coming from inside your tank this could indicate overheating

SHERMAN TURNER

and very dangerous pressure buildup. You should call a plumber or service professional immediately.

Patch burst pipes, if the water pipe freezes and breaks, your priority may be getting it working again, at whatever it takes get the job done. Just remember you are getting a temporary patch job, you still need the pipe or pipes fixed permanently.

Kitchen sinks come in many stylish designs now, but you can get basic stainless-steel sinks for around $100-$200. A higher quality stainless steel sink has a higher thicker gauge steel in a higher amount of nickel alloy and surface finish.

The high-quality stainless-steel sink should stay bright for many years and higher quality may cost from $200-$400.

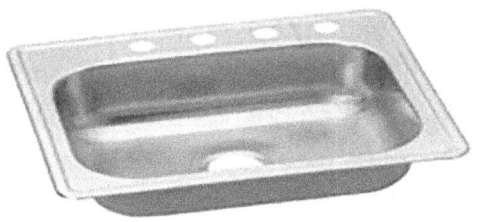

PLUMBING 'ESTIMATOR'

Cast-iron sinks, with enamel finish are very popular. It is possible to chip them, but they have very strong and very durable. Most cast-iron sinks are self-rimming, meaning they have raised lips they rest on top of the countertop; some are available with flush fit rims.

Types of shower almost nothing is more invigorating than the nice shower. If you have a good showerhead that you use. The type of showerhead you use will have a huge impact on the quality of your shower. There have been many advances in plumbing fixtures, so that your shower heated options are virtually unlimited.

Think showerhead is are fixed directly on the shower wall cannot be removed. However, some fixed chalets are adjustable and can be moved or aimed in different directions, which can be convenient when tall and short people live in the same house share the same shower.

Handheld showerheads are connected to a flexible hose that is mounted on the shower wall. With handheld showerheads, you can remove the shower. The attached hose usually allows for greater range of motion, making this the perfect showerhead for bathing pets, clean in the shower stall and hand washing your clothes.

Low-flow shower heads many consumers are turning to low-flow showerheads to reduce energy costs. These showerheads can be fixed or handheld and greatly reduce the amount of water that is sprayed from the nozzle.

The bathtub uses something called a "trip lever" as the waste system on the tub. There is also a P-trap connected to the trip lever assembly. Most of the time the P-trap will not be clogged or plugged up.

The drain will just be full of hair and can be cleaned out in several ways. How much hair and debris that is wrapped around the tub trip lever. Once you have cleaned all the visible here and

PLUMBING 'ESTIMATOR'

debris on the trip lever and drain, apply parts
with a light coating of grease.

Take a small hand snake or small electric snake
and run 2 or 3 feet of cable to make sure that
there are no obstructions in the drain line.

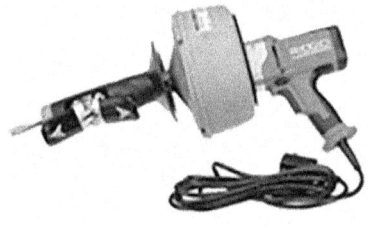

SHERMAN TURNER

Plumber 'Speaks' Q and A

Part Two

PLUMBING 'ESTIMATOR'

SHERMAN TURNER

Plumber 'Speaks'

Q and A

What is waterproofing? Where should waterproofing materials be used?

Use waterproofing sealers and paints which are recommended materials for their usage only and yes follow manufactures suggestions when using. Note: interior waterproofing sealers and paints will not alleviate all potential sources of basement leaking problem areas.

How often should downspouts and gutters be cleaned?

Rain gutters and downspouts should be cleaned regularly. Remove all leaves, dirt, and debris at least once or twice a year.

What are "French Drains?" Where should they

PLUMBING 'ESTIMATOR'

be used?

French Drains are used to channel water out of a building, which then would be pumped out of the building by way of a sump pump and sump pump pit. French drains also are a drain that accepts water that is drawn away and seeps into a gravel or dirt pit that you have not pumped.

What is the best and cheapest way to seal a leaking concrete wall?

Often, professionals applying waterproofing for concrete walls use fast setting "hydraulic cement."

For sealing concrete walls and waterproofing, first chisel a U- or V-shaped groove along the cracked wall. Next, clean out the groove, then fill along the crack with hydraulic cement.

What type of roof is best in an area of heavy rainfall?

Houses situated where heavy rainfall occurs each year usually have slate-type roofs. This is also not a requirement by building or plumbing inspectors.

What if your lawn is flat or sloping toward the house?

SHERMAN TURNER

A lawn that slopes toward your home can cause an accumulation of water around your basement walls resulting in leakage problems.

To correct this situation, slope and grade the lawn away from the house, if possible. If not possible, install an alternate method system such as a dry-well system or an approved storm drainage system.

Who uses concrete roofs?

It is very common to find commercial and residential buildings with concrete roofs.

How often should roofs be inspected?

Roofing systems should be inspected at least once a year.

Who best installs new roofs? A general contractor or a roofing contractor?

Hire a licensed roofing contractor who offers a guarantee for work and materials.

What is the correct wax seal for a toilet flange that is flush with the floor?

PLUMBING 'ESTIMATOR'

Try the Ultra-Seal-type wax gasket. I usually use the thickest wax ring I can find with the plastic flange (funnel-type pipe) embedded in it.

How do you measure the water evaporation from your swimming pool?

First, mark the water level on the wall of the pool. After some time, mark the water level again; it should be lower than the original mark.

Measure the area of the pool and then multiply that measurement by the difference of the two water levels. The result is the amount of water evaporated.

It is important to remember to turn off the pool's auto fill for this process. If you keep the auto fill on, the pool level will not drop much, if at all.

What causes water pipes to rattle, shake and make loud noisy sounds? How do you prevent it?

You should have water hammer arrestors installed, and water lines should be supported with the proper type of insulation as required per plumbing codes.

Should a toilet closet flange fit into or over the PVC drainpipe?

54

It can fit both ways. Just make sure it fits snug before you glue it. There are several types: a 3x4 flange fits over a 3-inch pipe and inside a 4-inch pipe, but there are also ones that will go over 4-inch pipe and inside 3-inch pipe.

Why would a toilet bowl go dry?

Like any fixture, a toilet can lose its seal (water) due to negative pressure or positive pressure or back siphonage. If the toilet is not vented properly.

Sometimes the vent on the roof gets plugged (bird's nest) and that could also be a factor.

Usually, it is because of an improper initial installation of a vent.

What would cause the hot water to be rusty and brown?

Sometimes heating water causes rust to get into the water lines, or a change in water pressure will cause rust in pipes to loosen and come through into the bathtub or other appliances.

If only the hot water is rusty, it is your hot water tank! The reason you're not seeing it in your cold water is because your cold-water lines run directly to your water fixtures.

PLUMBING 'ESTIMATOR'

What is the best way to move the toilet flange a couple of feet, and is there an adapter to go from cast iron to PVC?

Cut the toilet line down at the first 90-degree elbow from the flange and relocate it so that your toilet is still vented from its original vent. If it is dry-vented, you need to move the toilet and the vent. A no-hub band or Fernco brand coupling is the preferred mode of transition from cast iron to PVC.

What would cause a sink drain to sound like it is constantly bubbling and gurgling?

Gurgling noises in drains are normally the result of air being trapped in front of the body of water moving down the pipe (in no-vent piping). When pressure builds up, the air forces its way past the water, gurgling as it goes.

To stop the noise, install a vent in the pipe downstream from where the problem is.

When roughing in a new toilet, what is the correct distance from the center of the drain to the wall?

A minimum of sixteen inches off the finished side wall and a minimum of twelve inches of the finished back wall. However, an eighteen-inch side clearance is the international standard for wheelchair and assisted toilet maneuvering room.

What causes the popping or tinkling sounds as hot metal pipes cool off?

Popping sounds are caused by contractions, though the sounds are usually noisier when pipes are expanding as hot water runs through them. Expansion and contraction of the metal occurs when pipes are heated or cooled.

How can you increase the water pressure of a faucet?

Usually the aerator is plugged; but sometimes the rubber inside the stops under the sink wear out and tear away and plug up the end of the stops or the supply lines. Sometimes the supply lines get kinked and need to be replaced. If you get a flex tube to connect to your stop under the sink, you could attach it and blow the water into a bucket to see if you have pressure there. If there is pressure, then you know the problem is above the stops.

What is a plumber's snake used for?

It is a tool inserted into pipes to clean them if they are clogged. It can be operated manually or electric-powered.

PLUMBING 'ESTIMATOR'

Do you need to install a shower pan or liner when installing a tile shower on a concrete floor?

It's a good idea to install a shower pan so that water doesn't seep through and run into the rest of the basement. It's important to know that concrete is not waterproof. A waterproof membrane specific to the tile trade must be used. The installation guidelines are covered by the manufacturer and by local building plumbing codes.

What would cause water to seep out from under the toilet bowl onto the floor when the toilet is flushed?

You may have a partial clog somewhere in your pipes; therefore, the water is backing up in the first place. Perhaps the wax seal under the bottom of your toilet is broken. (You wouldn't be able to see it because it's **UNDER** the toilet.)

Can you increase water pressure from a gravity tank by increasing pipe sizes?

You cannot increase the pressure from a gravity tank by changing the pipe size, only the volume of water that will flow in a given period of time. The only way to increase the pressure is to increase the difference between the elevation of the tank and the elevation of the spigot.

Many people confuse an increase in water volume with an increase in water pressure because, in both cases, more water will flow in each period of time.

Increasing the pipe size can reduce elevation loss due to friction, which can translate in layman's terms as an "increase in pressure."

Why does the base of a toilet discolor?

The toilet is a ceramic bowl of standing water and any minerals or various chemicals that are in the water will "settle" in the pores of the ceramic over time.

How long does a Roto-Rooter (sewer) job last?

Depending on what's wrong with your sewer, a sewer repair job could take anywhere from 10 minutes to a whole day. Usually blockages in the main occur either from tree roots or breakages in the pipe.

If the line has been snaked and you still have a problem, I would recommend having the sewer, camera for pictures to find out what the problem and reason why the interior sewer system piping is not flowing correctly.

Why would a basement toilet suddenly go dry

PLUMBING 'ESTIMATOR'

and how can this be fixed?

If a toilet goes dry, it is probably not properly vented. When your sewer line is in use, the waste can actually create a siphon that can pull the water out of the trap in your toilet. The only way to fix it is to install the proper vent piping for the toilet.

Why would there be a sewer gas smell in the basement whenever the downstairs toilet is flushed?

Sometimes all the water evaporates out of the closest floor drain, usually in the basement. This would allow sewer gas smells into the house.

How do you keep tree/plant roots out of the main sewer line?

To keep roots from infiltrating the main sewer line, the only thing you can really do is snake out the sewer drainage system and get it as clear as possible. Then you can dump copper sulfate down the system.

It probably won't kill all the roots, but it will stop them from growing. This works best if you do it in the spring during the growing season. You should probably do it about twice a year in spring and in the fall. Another chemical to use is **"ROOTX"** because copper sulfate is illegal in some states.

Remember, putting any kind of chemicals down your sewer drain line is illegal in most countries around the world; it is considered an environmental hazard. If you are caught by authorities, you may be in serious trouble and the implications will be severe.

Therefore, it may be best to avoid using chemicals and instead rod or snake your sewer line, then cut down brush or trees growing over and around your sewer. Or, replace your sewer line and then maintain it by rodding it every six months.

How could you lose all hot water pressure without losing any cold water?

Lost water pressure (with retention of cold water) can be caused by some other simultaneous user of hot water, such as the clothes washer or dishwasher. These appliances drain or take all the hot water available and so your shower goes cold.

What causes a loud knocking sound like a jackhammer in the home plumbing when water is being shut off?

PLUMBING 'ESTIMATOR'

The knocking sound when shutting off water is called "water hammer," and you need to install "water hammer arrestors" in the piping system to prevent it.

Does a shower need a trap, and how and where is the trap located?

Yes, a shower does need a trap and sometimes a trap-refill, too! In most cases a shower trap is in the floor just below shower drain.

What company do you contract to remove asbestos?

Hire an asbestos abatement contractor, listed by or with the approval of "epa.gov." The Environmental Protection Agency (EPA) is the sanctioning governing body concerning asbestos type work.

Can you remove or get scratches out of the inside bowl of a toilet?

Yes, you can get your toilet refinished. Another option is to buy some porcelain patch and do it yourself.

What causes washing machine water to back up into the bathtub or toilet?

SHERMAN TURNER

The main drain line/sewer line under the house is clogged up. Have the main sewer Roto-Rootered and cleaned.

What is wrong when pipes make sound as you turn on the water?

If your water pipes are shaking making sounds when you turn on your water, the water pressure may be too high, and the water flow is causing water hammer on the pipes.

To alleviate this problem, you can lower your water pressure or install shock arresters in problematic locations. The noise may also be caused by a loose washer on a faucet.

Why does a shower drain get calcium buildup and a bathtub drain does not?

Most likely your shower floor is tiled, and the water is eroding your grout causing a buildup at the water line. Most bathtubs only have tile surrounding walls, so the water doesn't get blocked up to erode the grout.

How would you install plumbing such as a sink or toilet in a basement in which the drainage pipe exits high above the floor?

You should install a tank and a sewage ejector pump for this type of drainage. This system should

PLUMBING 'ESTIMATOR'

and must be approved by local licensed authorities prior to installing.

What does it mean that water is "metered?"

Metered water is the amount of water used and measured by the local utility company. When your water is metered, you pay for what you use or consume. This water is measured in meters, gallons, or feet.

How much slope or pitch do you need for a drainpipe to work properly?

If you don't have at least one-eighth of an inch of "slope" or "pitch," your drain will not work properly and may get clogged up regularly.

What does gray and brown water mean?

The color gray usually indicates dirty water (but not toilet water) and is usually seen in backed-up drainage from sinks, bathtubs, and showers.

Black or brown water is sewage water from the toilet.

Why does water gurgle and come out of the kitchen sink when a toilet is flushed?

If water gurgles when the toilet is flushed, there is—or may be—an air pocket in one of your pipes, or you could have a small clogging backup building.

Can you glue "PVC" and "CPVC" without primer?
Yes, you can. You should scratch the surface or use a piece of sandpaper on it to break the surface coating and clean.

What is the standard height for a shower head connection?

The standard height is seventy-eight inches, but in most new installations the height is now eighty inches.

What causes sink drains to make sucking sounds?

The sucking noise is a good thing, it means that the drain is working properly, and the noise comes from a swirl in the water going counterclockwise as it's drawn away. The noise is only air.

How do you know if you have a gas leak?

By using air pressure testing for gas leaks, then proceed to use soapy water to find the gas leak in the piping system. If the soapy water starts to bubble, there may be a leak.

PLUMBING 'ESTIMATOR'

How do you stop pipes from sweating?

To stop pipes from sweating, wrap them in insulation. Get insulation wherein the inside diameter of the insulation is larger than the outside diameter of the piping system.

The wall thickness of the insulation should be at least one inch. Make sure to use duct tape or electrical tape when covering turns in the piping system.

Why would kitchen sink or washroom sinks have a bad odor?

Most likely the P-trap or U-bend underneath a fixture-type sink has lost its trap seal due to evaporation. Refill the trap seal and pour water into the sinks to have a proper trap seal.

How do you keep a noisy hot water heater from making noise?

You can reduce the noise by draining the water heater and removing the lime deposits the best you can. Most likely you can't reduce the noise if the water heater is powered by gas and you have hard water.

The only way to fix this is to buy a new water heater and install a water softener

SHERMAN TURNER

Plumber 'Speaks' Plumbing Tips

Part Three

PLUMBING ESTIMATOR.

SHERMAN TURNER

Plumbing Tips

This very special section of 'Plumber Speaks' details various everyday situations that most homeowners will face when performing plumbing repairs.

By popular request, I have also addressed minor everyday issues, too. If these minor issues are not addressed, they may become major issues, which can result in a loss of hundreds even thousands of dollars in damage.

Knowing and remembering these *Special Tips* can save you 10% to 50% on your plumbing bills.

Faucets and Drains

These are the parts of your plumbing system that are most likely to break down and, therefore, will need repairs more often. Faucets leak or drip and drains get clogged up. When you add these repairs to fixing toilet problems, you have covered almost everything.

PLUMBING 'ESTIMATOR'

Toilets

Clogged toilets are the most common plumbing problem. If a toilet overflows or flushes sluggishly, clear the clog or backup with a plunger or closet auger. If the clog or backup persists, the problem may be in the main waste and vent system piping.

Puddle of Water or WaterLeak

Recurring water leakage on the floor around the toilet may be caused by a crack in the base of the toilet. This could also be coming from the toilet tank. It is best to check all water connections to the toilet.

Toilet Stability

If the toilet fixture shakes or rocks, then it is not fastened securely. Check the closet flange and secure all loose connections. If loose, replace the closet wax ring gasket with bolts and nuts.

Water on the floor around the toilet tank

Check to see if toilet tank is cracked and leaking. Also, check all water connections to the toilet fixture. You may have to tighten the tank bolts and then check all water connections. Sometimes you may have to insulate the toilet tank to prevent water condensation when very cold water enters a heated room.

Sink Drains

Every sink has a drain trap and a fixture drain. A sink gets clogged or plugged because of a build-up of soap and hair in the trap or fixture drain line. Remove clogs by using a

plunger or by disconnecting and cleaning the trap or by using a good hand-auger.

Clogs and Plug-ups

Clogs in sinks can be cleared with a plunger. Remove the pop-up drain plug and strainer on the sink first, and then plug off the overflow hole in the sink fixture by stuffing a wet rag into it, allowing air pressure to build. The air pressure while plunging will blow the obstruction free in the piping. Dump out the debris after removing the trap bend. You may need a small wire brush when cleaning the trap bend. Be sure to tighten all slip nuts.

Faucets

Just about all faucets develop leaks and drips. Repairs can be accomplished by replacing the mechanical parts inside the faucet body. The main thing is to figure out which kind and type of faucet you have and know the "make" of its parts.

PLUMBING 'ESTIMATOR'

If your old faucet continues to leak after repairs are made, then you know you will have to replace the old faucet with a new one.

Water Pressure

It is problematic when the water pressure at a faucet seems low or the water flow is partially blocked. Clean the faucet aerator. If this does not correct the situation, take a closer look under the sink. Check to see if the piping is old galvanized piping; if it is, then replace the corroded galvanized pipes with new copper or plastic type water piping.

Dishwashers

A dishwasher that is past its prime may be inefficient in many ways. It probably was not designed to be very efficient back in those days to begin with. If your dishwasher no longer cleans effectively, you are wasting time and hot water pre-rinsing the dishes. This alone can consume more energy and water than a complete wash cycle on a newer dishwashing machine. Therefore, even if the old dishwasher still runs, replacing it with an efficient new model is a good investment.

In terms of sizing and utility hookups, dishwashers are generally quite standard. If your old machine is a built-in model and your countertops and cabinets are standard size, most full-size dishwashers will fit right in.

Of course, you should always measure the dimensions of the old unit before shopping for a new one to avoid any unpleasant surprise at installation time. Also, be sure to review and follow all manufacturers' work and installation instructions.

Replacing an Old Inefficient Dishwasher.

This is a pretty straight forward project that usually takes just a few hours. Your energy savings begin with the first load of dishes and continue with every load thereafter. That is why I always recommend using a newer model dishwasher appliance.

Food Disposers

These are standard equipment in modern homes and most of us have come to depend on them to macerate our plate leavings so our crumbs can exit the house along with wastewater from the sink drain, for efficient disposal of organic waste.

If your existing disposer needs to be replaced you will find the job is very easy, especially if you select a replacement appliance that is the same

PLUMBING 'ESTIMATOR'

model as the old one. Now you can reuse the same existing mounting assembly and drain plumbing.

Each Appliance has a power rating between 1/3 and 1 HP (horsepower). The more powerful models bog down less under heavy use and the motors last longer because they do not have to work as hard. They are more costly too.

Disposers are hardwired to a switch mounted in an electric box in the wall above the countertop. The actual electrical hookup of the appliance is quite simple (you only must join two wires) but hire an electrician, it's best to be safe.

Icemakers

The most expensive refrigerators come with icemakers as standard equipment and practically every model features them as an option. (A refrigerator with an icemaker usually cost about $100 - $200 more than one without.)

Most icemakers come either pre-installed or are purchased as an accessory when you buy your new refrigerator.

An icemaker receives its supply of water for making cubes through a ¼-inch supply line that runs from the icemaker and connects to a water pipe. The supply line runs through a valve in the refrigerator and is controlled by the solenoid valve that monitors the water supply and sends the water into the icemaker itself, where it is turned into ice cubes.

Then the cubes drop down into a bin. As the ice levels rise, this will raise a bail wire that is connected to water shutoff valve. When the bin gets full, the bail wire triggers a mechanism that shuts off the water supply valve. Unfortunately, water supply lines and the connection points are very common sources of water leaks.

Automatic Icemakers

These are simple to install as long as your refrigerator is icemaker ready. Make sure to buy the correct model for your appliance.

Standpipe Drains

In many houses the washing machine drain hose is hung loosely over the side of the utility sink. This type of installation is now not allowed by many buildings and plumbing codes.

Most codes require homeowners to install a standpipe drain that allows the washing machine to drain into the utility sink's drain line piping system. Using a 2-inch standpipe is mostly

PLUMBING 'ESTIMATOR'

required by most codes with a 2-inch P-trap. The top of the standpipe should be higher than the highest level in the washing machine.

Sump Pumps

Prices range from about $150 to $600 or more, depending on the quality and features of a sump pump. First, decide between a pedestal and submersible pump. Which one is best for your purpose and job?

A submersible pump remains out of sight, sits in water a good deal of the time, and has a lifespan of ten years. Most manufactures offer a limited one- to five-year guarantee.

A submersible pump is safe and can be an important advantage if the basement is used as a primary living area. This pump is much safer if there are children in the house.

Sump pump

Is measured by horsepower, which ranges from 1/6 to ½ HP. However, more important is the number of gallons per minute (GPM) a pump will move.

This capacity factor of both the bottom of the pump's efficiency and the "head" or "lift" is the vertical distance from the bottom of the sump to the highest point of discharge. Also, note whether the pump is strong enough to pass small solids, such as leaves and twigs.

Switches on Sump Pumps

These come in several types including floats, diaphragms, and mercury switches. It does not really matter which type you choose but be sure it's automatic.

Buy Cast Iron Sump Pumps

Last longer than plastic or iron type pumps because they don't break easily. Make sure the power cord is long enough because electrical extension cords are not to be used on sump pumps.

Sump Pump installations

Always place "pavers" or "bricks" underneath the sump pump so mud, dirt and grind does not plug up the operation of the pump and motor.

Clogged Sewers.

A clogged sewer is a problem that requires immediate attention. There are some simple techniques to clearing a clogged or plugged up sewer line. Chemical cleaners can effectively clear out any organic obstruction by dissolving

PLUMBING 'ESTIMATOR'

them. The chemicals are generally available in hardware stores. Be sure to follow the manufactures directions carefully.

Use a "plumber's plunger" which is about 10 or 12 feet long, to generate enough pressure to blow-out obstructions quite quickly

Plumbers and Clogged Sewers

Usually use this at the house trap and fresh-air-inlet outside the house. Use an electric snake when and if the obstruction is in the form of tree roots, instead of common obstructions like toilet paper, rags, leaves and garbage. You may then need the use of an electric snake tool.

For extreme cases, call in a professional Plumbing Company who may use a camera to find the clog and then use a Jet Hydrovac Machine to disperse it.

Water Heater:

Makes strange noises. When the hot water tank is heating water, the expanding and contracting metal parts; or more likely, minerals and hard water scale build-ups these accumulations inside the tank. This can cause the noises coming from your water heater.

To avoid scale buildup, every few months, open the drain valve at the base of the tank, then flush out the tank until all the rust and scales runs out and you see clear water.

Hot Water Tank manufacturers recommend draining and flushing your water heater tank every six months in hard water areas. This helps remove sediment and minerals that build-up at the tank bottom.

If you hear boiling sounds inside your hot water tank, this could indicate overheating and very dangerous pressure build-up. Call a professional plumber immediately.

Banging, gurgling, or popping sounds are caused by the build-up of hard water sediment heating up and exploding inside the tank.

Types of Showerheads

The type of showerhead you use will have a huge impact on the quality of your shower. There have been many advances in plumbing, so your choice and options for a showerhead are virtually unlimited. A handheld showerhead is connected to a flexible hose that is mounted on the shower wall. With handheld showerheads, you can remove the showerhead and replace it easily.

PLUMBING 'ESTIMATOR'

The attached hose usually allows for greater range of motion.

Low-flow showerheads reduce energy costs. These showerheads can be fixed or handheld and greatly reduce the amount of water that is sprayed from the nozzle, allowing for less wasted water.

Kitchen Sinks

These come in many stylish designs now. You can get a basic stainless-steel sink for around $110 to $160. A higher-quality stainless steel sink has a higher (thicker) gauge of steel and a higher amount of nickel alloy and surface finish, but with a much higher cost.

Cast Iron Sinks

With an enamel finish, cast iron sinks are very popular. It is possible to chip them, but they are very strong and very durable. Most cast iron sinks are self-rimming, and some are available with flush-fit rims.

Bathtubs

As one of the most used plumbing fixtures in the house, your bathtub is often covered with soap residue. Your bathtub will eventually get plugged up with hair and dirt and its drain will gradually become clogged.

Older Toilets

When water in an old toilet keeps running and filling the toilet tank, try replacing the old fill water valve inside the toilet tank.

What purpose does Vent Piping serve?

In the Plumbing System. The vent piping system should keep sewer gases from entering your house and help your plumbing drainage piping drain properly

PLUMBING 'ESTIMATOR'

SHERMAN TURNER

Budget Cost Estimating

PLUMBING 'ESTIMATOR'

Chapter #9

Budget Cost Estimating

The **Author**, Master / Plumber has written this book specifically benefit **Homeowners** and **Small Businesses** who face those tough decisions trying to save money, because upkeep and repairs are costly.

This book takes you by the hand and helps explain the *do's* and *don'ts*. Most plumbing books say *do this* but never say *don't do that*. Confusing isn't it!

The most important reason why I wrote this book to help fight against unscrupulous **Contractors**. Who sometimes are always ripping-off the elderly and the disabled!

This book gives **Homeowners** and **Small Businesses** an opportunity to evaluate Contractors prices and scope of work.

PLUMBING 'ESTIMATOR'

Cost estimating is a well-developed discipline. By understanding cost estimating and using standard estimation techniques, you can improve your forecasts. The book (*Plumber*) is complete guide to project cost estimating! Computer charts will walk you through the key concepts and major estimating techniques.

Today everyone has a **Computer**. To track and estimate costs, you need estimating software program. As an expert estimate of 40 years I have developed computed labor and cost Charts to easily set-up in your **Computers**.

Using my free Charts, labor rates equal hours times the labor hourly costs, gives the estimated costs. I paid over $10K for the software. I am too happy to give and save **Homeowners** and **Small Businesses** savings of $10K or more.

SHERMAN TURNER

What are different categories of **estimating**?

- Budget
- Planning
- Feasibility

Students can easily use free Charts, labor rates equal hours times the labor hourly costs, gives the estimated costs.

PLUMBING 'ESTIMATOR'

Estimating Piping & Fittings Labor

<u>Notice:</u> All labor rates are reference as average, depending on different job material applications. Labor rates should be adjusted accordingly.

Labor = parts of an hour = .10 =   6 minutes
Labor = parts of an hour = .25 = 15 minutes
Labor = parts of an hour = .5 =   30 minutes
Labor = parts of an hour = .75 = 45 minutes
Labor = parts of an hour = 1.0 = 60 minutes

DWV Copper Pipe Size	Pipe Labor	Fitting Labor
1 1/2	0.01	0.1
2	0.01	0.1
3	0.03	0.15
4	0.04	0.25
6	0.06	0.5

SHERMAN TURNER

Estimating Piping & Fittings Labor

Notice: All labor rates are reference as average, depending on different job material applications. Labor rates should be adjusted accordingly.

Labor = parts of an hour = .10 = 6 minutes
Labor = parts of an hour = .25 = 15 minutes
Labor = parts of an hour = .5 = 30 minutes
Labor = parts of an hour = .75 = 45 minutes
Labor = parts of an hour = 1.0 = 60 minutes

Copper "L" Pipe Size	Pipe Labor	Fitting Labor
1/2	0.01	0.1
3/4	0.01	0.1
1	0.01	0.1
1 1/4	0.02	0.15
1 1/2	0.02	0.15
2	0.02	0.25
3	0.03	0.35
4	0.04	0.5

PLUMBING 'ESTIMATOR'

Estimating Piping & Fittings Labor

Notice: All labor rates are reference as average, depending on different job material applications. Labor rates should be adjusted accordingly.

Labor = parts of an hour = .10 = 6 minutes
Labor = parts of an hour = .25 = 15 minutes
Labor = parts of an hour = .5 = 30 minutes
Labor = parts of an hour = .75 = 45 minutes
Labor = parts of an hour = 1.0 = 60 minutes

PVC #40 Pipe Size	Pipe Labor	Fitting Labor
1/2	0	0.03
3/4	0	0.03
1	0	0.03
1 1/4	0.01	0.03
1 1/2	0.01	0.03
2	0.02	0.03
3	0.03	0.05
4	0.03	0.05

SHERMAN TURNER

Estimating Piping & Fittings Labor

Notice: All labor rates are reference as average, depending on different job material applications. Labor rates should be adjusted accordingly.

Labor = parts of an hour = .10 = 6 minutes
Labor = parts of an hour = .25 = 15 minutes
Labor = parts of an hour = .5 = 30 minutes
Labor = parts of an hour = .75 = 45 minutes
Labor = parts of an hour = 1.0 = 60 minutes

PVC Plastic Pipe Size	Pipe Labor	Fitting Labor
1 1/2	0.02	0.05
2	0.02	0.05
3	0.04	0.05
4	0.05	0.05
6	0.07	0.05
8	0.1	0.1
10	0.15	0.1
12	0.2	0.1

PLUMBING 'ESTIMATOR'

Estimating Piping & Fittings Labor

Notice: All labor rates are reference as average, depending on different job material applications. Labor rates should be adjusted accordingly.

Labor = parts of an hour = .10 = 6 minutes
Labor = parts of an hour = .25 = 15 minutes
Labor = parts of an hour = .5 = 30 minutes
Labor = parts of an hour = .75 = 45 minutes
Labor = parts of an hour = 1.0 = 60 minutes

Black Steel Pipe Size	Pipe Labor	Fitting Labor
1/2	0.01	0.1
3/4	0.01	0.1
1	0.02	0.1
1 1/4	0.02	0.15
1 1/2	0.02	0.15
2	0.03	0.25
3	0.04	0.35
4	0.06	0.5

SHERMAN TURNER

Estimating Piping & Fittings Labor

Notice: All labor rates are reference as average, depending on different job material applications. Labor rates should be adjusted accordingly.

Labor = parts of an hour = .10 =   6 minutes
Labor = parts of an hour = .25 = 15 minutes
Labor = parts of an hour = .5 =   30 minutes
Labor = parts of an hour = .75 = 45 minutes
Labor = parts of an hour = 1.0 = 60 minutes

No-Hub CI Pipe Size	Pipe Labor	Fitting Labor
1 1/2	0.01	0.1
2	0.01	0.1
3	0.02	0.15
4	0.03	0.25
6	0.04	0.35
8	0.06	0.5
10	0.08	0.75
12	0.1	1

PLUMBING 'ESTIMATOR'

Estimating Piping & Fittings Labor

Notice: All labor rates are reference as average, depending on different job material applications. Labor rates should be adjusted accordingly.

Labor = parts of an hour = .10 = 6 minutes
Labor = parts of an hour = .25 = 15 minutes
Labor = parts of an hour = .5 = 30 minutes
Labor = parts of an hour = .75 = 45 minutes
Labor = parts of an hour = 1.0 = 60 minutes

Galv. Steel Support Size	Pipe Labor	Fitting Labor
1 1/2	0.01	0.1
2	0.01	0.1
3	0.02	0.15
4	0.03	0.15
6	0.04	0.25
8	0.06	0.25
10	0.08	0.35
12	0.1	0.35

SHERMAN TURNER

Estimating Piping & Fittings Labor

Notice: All labor rates are reference as average, depending on different job material applications. Labor rates should be adjusted accordingly.

Labor = parts of an hour = .10 =   6 minutes
Labor = parts of an hour = .25 = 15 minutes
Labor = parts of an hour = .5 =   30 minutes
Labor = parts of an hour = .75 = 45 minutes
Labor = parts of an hour = 1.0 = 60 minutes

Copper Type Support Size	Pipe Labor	Fitting Labor
1 1/2	0.01	0.1
2	0.01	0.1
3	0.02	0.15
4	0.03	0.15
6	0.04	0.25
8	0.06	0.25
10	0.08	0.35
12	0.1	0.35

PLUMBING 'ESTIMATOR'

Estimating Piping & Fittings Labor

Notice: All labor rates are reference as average, depending on different job material applications. Labor rates should be adjusted accordingly.

Labor = parts of an hour = .10 =   6 minutes
Labor = parts of an hour = .25 = 15 minutes
Labor = parts of an hour = .5 =   30 minutes
Labor = parts of an hour = .75 = 45 minutes
Labor = parts of an hour = 1.0 = 60 minutes

Steel Hangers Pipe Size	Pipe Labor	Fitting Labor
1/2	0	0.1
3/4	0	0.1
1	0	0.1
1 1/4	0.01	0.15
1 1/2	0.01	0.15
2	0.02	0.15
3	0.03	0.25
4	0.03	0.25

SHERMAN TURNER

Estimating Piping & Fittings Labor

Notice: All labor rates are reference as average, depending on different job material applications. Labor rates should be adjusted accordingly.

Labor = parts of an hour = .10 =   6 minutes
Labor = parts of an hour = .25 = 15 minutes
Labor = parts of an hour = .5 =   30 minutes
Labor = parts of an hour = .75 = 45 minutes
Labor = parts of an hour = 1.0 = 60 minutes

Cop. Hangers Pipe Size	Pipe Labor	Fitting Labor
1/2	0	0.1
3/4	0	0.1
1	0	0.1
1 1/4	0.01	0.15
1 1/2	0.01	0.15
2	0.02	0.15
3	0.03	0.25
4	0.03	0.25

PLUMBING 'ESTIMATOR'

Estimating Piping & Fittings Labor

<u>Notice</u>: All labor rates are reference as average, depending on different job material applications. Labor rates should be adjusted accordingly.

Labor = parts of an hour = .10 =   6 minutes
Labor = parts of an hour = .25 = 15 minutes
Labor = parts of an hour = .5 =   30 minutes
Labor = parts of an hour = .75 = 45 minutes
Labor = parts of an hour = 1.0 = 60 minutes

Split Ring Galv. Steel	Pipe Labor	Fitting Labor
1/2	0	0.05
3/4	0	0.05
1	0	0.05
1 1/4	0.01	0.1
1 1/2	0.01	0.1
2	0.02	0.1
3	0.03	0.15
4	0.03	0.15

SHERMAN TURNER

Estimating Piping & Fittings Labor

Notice: All labor rates are reference as average, depending on different job material applications. Labor rates should be adjusted accordingly.

Labor = parts of an hour = .10 = 6 minutes
Labor = parts of an hour = .25 = 15 minutes
Labor = parts of an hour = .5 = 30 minutes
Labor = parts of an hour = .75 = 45 minutes
Labor = parts of an hour = 1.0 = 60 minutes

Split Ring Copper	Pipe Labor	Fitting Labor
1/2	0	0.05
3/4	0	0.05
1	0	0.05
1 1/4	0.01	0.1
1 1/2	0.01	0.1
2	0.02	0.1
3	0.03	0.15
4	0.03	0.15

PLUMBING 'ESTIMATOR'

"DWV" Copper Pipe & Fittings

JOB =			
	DWV Copper Pipe & Fittings		
Item	Description	Count	Totals

SHERMAN TURNER

PEX Pipe & Fittings

JOB =			
PEX Pipe & Fittings			
Item	**Description**	**Count**	Totals

PLUMBING 'ESTIMATOR'

"L" Copper Pipe & Fittings

JOB =			
"L" Pipe & Copper Fittings			
Item	**Description**	**Count**	Totals

SHERMAN TURNER

IPS Black Steel and Fittings

JOB =			
	IPS Black Steel and Fittings		
Item	**Description**	**Count**	**Totals**

PLUMBING 'ESTIMATOR'

Gas Pipe and Fittings

JOB =			
Gas Pipe and Fittings			
Item	**Description**	**Count**	Totals

SHERMAN TURNER

Hangers and Miss. Items

JOB =			
Hangers and Miss. Items			
Item	Description	Count	Totals

PLUMBING 'ESTIMATOR'

Plumbing Material Takeoff Sheet

JOB =			
Plumbing Materials =			
Type	**Description**	**Count**	Totals

SHERMAN TURNER

PVC WATER

JOB =			
PVC WATER PIPE MATERIALS			
Item	**Description**	**Count**	**Totals**

PLUMBING 'ESTIMATOR'

PVC DRAIN PIPE

JOB =			
PVC DRAIN PIPE MATERIALS			
Item	Description	Count	Totals

SHERMAN TURNER

Plumbing Fixtures

JOB =			
Plumbing Fixtures			
Item	Description	Count	Totals

PLUMBING 'ESTIMATOR'

Plumbing Equipment

JOB =			
Plumbing Equipment =			
Type	**Description**	**Count**	Totals

SHERMAN TURNER

Sub-Contractors Takeoff Sheet

JOB =		
Sub-Contractors =		
System	**Description**	Totals
1		
2		
3		
4		
5		
6		

PLUMBING 'ESTIMATOR'

Plumbing Job Summary

JOB =			
Summary =			
Item	**Work or Job Description**	**Cost**	Totals

SHERMAN TURNER

Free Estimate and Evaluations

Master-Plumber and now Author, has more than 40 years' experience and more than 30 years' experience as a Senior Plumbing Estimator.

Thank you, for allowing me to be of any type plumbing service for your *"Free Plumbing Estimate Evaluation."*

A *"Free Estimate Evaluation"* is an appraisal of something to determine its worth or fitness. In most cases this eliminates Plumbing Contractor rip-offs!

Plus, *"Free Estimate Evaluation"* is an appraisal which recommends you make and stay within your budget. This is very helpful for many Job Projects and Jobs!

Please contact me at: plumb14@yahoo.com via email. Also please explain any related details in full. That way I can be more of service to you!
(Special) For same day 24-hour bids!

PLUMBING 'ESTIMATOR'

For **(SB)** Small Businesses and **(PC)** Plumbing Contractors: seeking win more competitive bids it is recommended to have an experienced estimate reviewer, review your *"Bid Summary!"* (1 - 4 pages) To make your bid a competitive winner! **Email via:** plumb14@yahoo.com

(Special) For same day 24-hour bids!

For **homeowners** and all **others,** I recommend an estimate budget evaluation cost! Because if you feel like Plumber price is too high, nine times out of ten, the Plumber's price is too high!

Most **Plumbers** and **Contractors** rip you off for ten times as much as the original cost. Fight back get an estimate budget evaluation cost! **Email via:** plumb14@yahoo.com

SHERMAN TURNER

"Notes"

PLUMBING 'ESTIMATOR'

"Notes"

SHERMAN TURNER

"Notes"

PLUMBING 'ESTIMATOR'

"Notes"

SHERMAN TURNER

"Contacts"

1# Address _____

Ph: _____Fax: _____

Name _____

2# Address _____

Ph: _____Fax: _____

Name _____

3# Address _____

Ph: _____Fax: _____

Name _____

4# Address _____

Ph: _____ Fax _____

Name _____

PLUMBING 'ESTIMATOR'

"Contacts"

1# Address _____

Ph: _____Fax: _____

Name _____

2# Address _____

Ph: _____Fax: _____

Name _____

3# Address _____

Ph: _____ Fax: _____

Name _____

4# Address _____

Ph: _____ Fax _____

Name _____

SHERMAN TURNER

"Contacts"

1# Address _____

Ph: _____Fax: _____

Name _____

2# Address _____

Ph: _____Fax: _____

Name _____

3# Address _____

Ph: _____Fax: _____

Name _____

4# Address _____

Ph: _____ Fax _____

Name _____

PLUMBING 'ESTIMATOR'

"Contacts"

1# Address _____

Ph: _____Fax: _____

Name _____

2# Address _____

Ph: _____Fax: _____

Name _____

3# Address _____

Ph: _____Fax: _____

Name _____

4# Address _____

Ph: _____ Fax _____

Name _____

www.ingramcontent.com/pod-product-compliance
Lightning Source LLC
Chambersburg PA
CBHW070646220526
45466CB00001B/320